Musical Education & Fitness Walk

Wanda Micheal

Archway Publishing books may be ordered through booksellers or by contacting:

Archway Publishing
1663 Liberty Drive
Bloomington, IN 47403
www.archwaypublishing.com
844-669-3957

ISBN: 978-1-6657-1606-2 (sc)
ISBN: 978-1-6657-1607-9 (e)

Library of Congress Control Number: 2021924814

Print information available on the last page.

Archway Publishing rev. date: 08/02/2022

Contents

Musical Education & Fitness Walk
Game Instructions

Choose a moderator to run the game, then pick your favorite genre of music to play. Use one piece of 8.5 x 11 copy paper for each participant in the game. Use a marker or pen to create a sheet for each participant and tape them to the floor in a circle. If more than twenty-five cards are needed, you can choose a question from the book, or create your own questions.

This game can be modified to fit any topic or curriculum, such as English, History, Social Studies, Foriegn Languages, Grammar, etc.

It can be used pedagogically for educators too, such as in topic review and test preparation.

It's also great for general trivia and parties for all ages!

The game begins when the music starts. Participants walk around the circle as the music plays. When the music stops participants move to the nearest sheet of paper. Each participant must then answer the question on the sheet they moved to, or follow the instructions on the sheet they are beside. If they answer a question incorrectly or land on sheet **2**, they are out. If they land on sheet **1** the moderator can choose a question or vocabulary word from the book or within their modified version of the game. Remove one sheet after each turn, except sheets **1** and **2**. The answers can be duplicated more than once during the game, and it is up to the moderator to determine if the answers not included in the book are correct. Keep going until there is a winner.

Tip: the best listener is most often the winner of the game.

Let's have some fun!!!

- Sheet 1-?
- Sheet 2-You're Out.
- Sheet 3-Do Ten Jumping Jacks.
- Sheet 4-The Trumpet Is a _____ Instrument.
- Sheet 5-Congruent Definition.
- Sheet 6-Jog in Place and Count to Ten.
- Sheet 7-Name a Woodwind Instrument.
- Sheet 8-Do Your Own Dance and Count to Ten.
- Sheet 9-Name a Famous Athlete.
- Sheet 10-Tempo Definition.
- Sheet 11-Name Two Science Vocabulary Words.
- Sheet 12-T or F? The Study of Viruses Is called _____?
- Sheet 13-The Violin Is a _____ Instrument.
- Sheet 14-Shout Your Name, Age, and School Loud and Proud!
- Sheet 15-T or F? An Octagon has six sides.
- Sheet 16-Is the Sun a Star?
- Sheet 17-Math Definition.
- Sheet 18-Name a Percussion Instrument.
- Sheet 19-Congruent Definition.
- Sheet 20-What Is the Moderator's Name?
- Sheet 21-Fitness Definition.
- Sheet 22-Shout 'I Am Amazing!'
- Sheet 23-Science Definition.
- Sheet 24-T or F? The Saxophone is a String Instrument.
- Sheet 25-Shout 'I Am Brilliant!'

Math Q&A

- Count to five.
- Count to ten.
- Count to one hundred by fives.
- Count to one hundred by tens.
- Count to twenty-five.
- Count to fifty.
- Count to five in Spanish. Answer: uno, dos, tres, cuatro, cinco.
- Count to ten in Spanish. Answer: uno, dos, tres, cuatro, cinco, seis, siete, ocho, nueve, diez.

Equations

- $3+1 = 4$
- $2+2 = 4$
- $10+5 = 15$
- $10+10 = 20$
- $10+2+3 = 15$
- $1+2+1 = 4$
- $5+5 = 10$
- $6+6 = 12$
- $50+50 = 100$
- $100+400 = 500$
- $8+8+8+8 = 32$
- $1+3+4+1 = 9$
- $500+400 = 900$
- $50+60-10 = 100$
- $20-10 = 10$
- $50-30 = 20$
- $12-6-2 = 4$
- $100-15 = 85$
- $8-3 = 5$
- $6-6 = 0$
- $18-15 = 3$
- $192-3 = 189$
- $55-6 = 49$
- $5-5 = 0$

- 12-6 = 6
- 20-5 = 15
- 150-49 = 101
- 5-3 = 2
- 30-15 = 15
- 100-98 = 2
- 10-5-2 = 3
- 9+9-3 = 15
- 50-40+10 = 20
- 25-5+2 = 22
- 2x2 = 4
- 5x3 = 15
- 8x8 = 64
- 4x4 = 16
- 9x9 = 81
- 1x2 = 2
- 9x8 = 72
- 5x5 = 25
- 10x2 = 20
- 3x8 = 24
- 4x3 = 12
- 7x7 = 49
- 6x6 = 36
- 2x6 = 12
- 6x7 = 42
- 3x3 = 9
- 20x3 = 60
- 50x3 = 150
- 8x7 = 56
- 10 divided by 5 = 2
- 8 divided by 4 = 2
- 72 divided by 9 = 8
- 64 divided by 8 = 8
- 49 divided by 7 = 7
- 100 divided by 2 = 50
- 4 divided by 2 = 2
- 81 divided by 9 = 9
- 9 divided by 3 = 3
- 12 divided by 2 = 6

- 18 divided by 9 = 2
- 56 divided by 8 = 7
- 6 divided by 3 = 2
- 50 divided by 2 = 25
- Five pennies = five cents.
- Five dimes = fifty cents.
- Two quarters = fifty cents.
- Two quarters, two dimes, two nickels, and five pennies = eighty-five cents.
- Five nickels = twenty-five cents.
- Round 21 to the nearest ten. Answer: 20
- Round 48 to the nearest ten. Answer: 50
- Round 54 to the nearest ten. Answer: 50
- Round 7 to the nearest ten. Answer: 10
- Round 125 to the nearest hundred. Answer: 100
- Round 456 to the nearest hundred. Answer: 500
- Round 899 to the nearest hundred. Answer: 900
- Round 238 to the nearest hundred. Answer: 200
- Round 1,021 to the nearest thousand. Answer: 1,000
- Round 5,601 to the nearest thousand. Answer: 6,000
- Round 8,456 to the nearest thousand. Answer: 8,000
- Round 9,999 to the nearest thousand. Answer: 10,000
- Round 34,555 to the nearest ten thousand. Answer: 30,000
- Round 87,333 to the nearest ten thousand. Answer: 90,000
- Round 23,555 to the nearest ten thousand. Answer: 20,000
- Round 95,444 to the nearest ten thousand. Answer: 10,0000
- Round 100,555 to the nearest hundred thousand. Answer: 200,000
- Round 325,301 to the nearest hundred thousand. Answer: 300,000
- Round 666,666 to the nearest hundred thousand. Answer: 700,000
- Round 999,999 to the nearest hundred thousand. Answer: 1,000,000
- How many sides does a rectangle have? Answer: four.
- How many sides does a hexagon have? Answer: six.
- How many sides does an octagon have? Answer: eight.
- How many square faces does a cube have? Answer: six.
- How many sides does a quadrilateral have? Answer: four.
- How many sides does a pentagon have? Answer: five.
- How many sides does a decagon have? Answer: ten.
- How many sides does a square have? Answer: four.
- A triangle with two sides of equal length is an_____. Answer: Isosceles triangle.
- A triangle with all sides of equal length is an_____. Answer: Equilateral triangle.

- A triangle with no sides of equal length is a _____. Answer: Scalene triangle.
- What are these numbers, "first, second, third, fourth, fifth," called? Answer: Ordinal numbers.
- Name the number in the tenths place in .0534. Answer: 0
- Name the number in the hundredths place in .5355. Answer: 3
- Name the number in the thousandths place 48.14678. Answer: 6
- Name the number in the ten thousandths place 5.55654. Answer: 5
- Reduce 2/4 to lowest terms as a proper fraction. Answer: 1/2
- Reduce 2/8 to lowest terms as a percentage. Answer: 25%
- A quarter past three is what time? Answer: 3:15
- What is another word for noon? Answers: Noontide, twelve noon, high-noon, midday, noonday.
- How many millimeters are in 1 meter? Answer: 1000 millimeters.
- How many centimeters are in 1 meter? Answer: 100 centimeters.
- 1 meter is equal to how many millimeters or how many centimeters? Answer: 1000 millimeters and 100 centimeters.
- Dallas gave her brothers two blue cars, three red cars, and four green cars. How many cars did she give them in all? Answer: Nine.
- Isaiah has five apples and three pears. Mason has three apples and five pears. Which one has the most pears? Answer: Mason.
- Madison has seven candles lit. Two blew out. How many candles are still lit? Answer: Five.

Science Q&A

- Why is the sky blue? Answer: Sunlight reaches Earth's atmosphere and is scattered in all directions by all the gasses and particles in the air. Blue light is scattered more than other colors because it travels in shorter, smaller waves. This is why we see a blue sky most of the time.
- Are dolphins mammals? Answer: Yes.
- What is a black hole? Answer: A region of spacetime where gravity is so strong that nothing can escape from it.
- How are rainbows made? Answer: When sunlight strikes raindrops in front of a viewer at a precise angle.
- How far away is the Earth from the sun? Answer: 93,000,000 million miles away.
- How do airplanes fly? Answer: When the movement of air across their wings creates an upward force on the wings that is greater than the force of gravity pulling the plane toward the Earth.
- How do flies walk on the ceiling? Answer: Each foot comes with a pair of claws used to grip tiny irregularities on rough surfaces.
- Are sharks mammals? Answer: Sharks are cold-blooded, whereas mammals are warm-blooded, so they are not mammals.
- What orbits around the Earth? Answer: The moon.
- Humans are classified as _____ Answer: Mammals.
- How many days are in a week? Answer: Seven.
- How many months are in a year? Answer: Twelve.
- What was the Manhattan Project? Answer: A research and development undertaking during World War II.
- What did the Manhattan Project develop? Answer: It produced the first nuclear weapons.
- Why shouldn't you mix chlorine bleach and ammonia? Answer: When combined, these two common household cleaners release toxic chloramine gas.
- What is a magnet? Answer: A material or object that produces a magnetic field.
- What does LCD stand for? Answer: Liquid-crystal display.
- What is a catalyst? Answer: A substance that can be added to a reaction to increase the reaction rate without getting consumed in the process.
- What is an atomic bomb? Answer: A nuclear weapon that explodes due to the extreme energy released by nuclear fission.
- What is a hydrogen bomb? Answer: A nuclear weapon that explodes from the intense energy released by nuclear fusion.
- What do you call the long, thin clouds forming behind airplanes as they fly? Answer: Contrails, short for condensation trails.
- What percentage of the cells in your body are human? Answer: 43%
- What is the biggest planet in our solar system? Answer: Jupiter.

- Does science change? Answer: The accepted views of science knowledge can change over time.
- What is matter? Answer: A physical substance that occupies space and possesses rest mass.
- What is mass? Answer: A large body of matter with no definite shape.
- What is a proton? Answer: A small particle that exists in the nucleus of every atom and has a positive charge of electricity.
- What is a neutron? Answer: A subatomic particle of about the same mass as a proton, but without an electric charge.
- What is a nucleus? Answer: The central part of an atom that comprises all the atomic mass, and consists of protons and neutrons.
- What is an atomic number? Answer: The number of protons found in the nucleus of each atom of that element.
- T or F? Oxygen is a chemical element with the symbol O and the atomic number 8. Answer: True.
- T or F? Hydrogen is the chemical element with the symbol H and the atomic number 1. Answer: True.
- T or F? Carbon is a chemical element with the symbol C and the atomic number 6: True.
- T or F? Chlorine is a chemical element with the symbol CL and the atomic number 17: True.
- T or F? Lithium (LI) is primarily used as a psychiatric medication to treat bipolar disorder and major depressive disorder along with antidepressants. Answer: True.
- What is another name for a tidal wave? Answers: Tsunami, giant wave, rogue wave, giant sea swell.
- Are dogs carnivores or omnivores? Answer: Omnivores, which are animals that eat both plants and animals.
- What is the periodic table? Answer: A tabular display of the chemical elements, which are arranged by atomic number, electron configuration, and recurring chemical properties.
- What is the seventh element on the periodic table of elements? Answer: Nitrogen.
- T or F? DNA is the acronym for 'Deoxyribonucleic acid.' Answer: True.
- What is the highest mountain on Earth? Answer: Mount Everest.
- What is the name of the closest star to the Earth? Answer: Sun.
- What is a cold-blooded animal? Answer: Animals that cannot regulate their internal body temperature with the change in the environment.
- What are warm blooded animals? Answer: Animals that can maintain a body temperature higher than their environment.
- T or F? Frogs are cold blooded animals. Answer: True.
- T or F? Warm blooded animals include birds and mammals. Answer: True.
- Pure water has a pH of ____? Answer: 7
- The molten rock that comes from an erupting volcano is called ____? Answer: Lava.
- What is the part of the human skeleton which protects our brain? Answers: Cranium or skull.
- T or F? The fastest land animal in the world is the zebra. Answer: False. It is a cheetah.
- Is the compound 'HCl' an acid or base? Answer: Acid.

- What famous scientist was awarded the Nobel Peace Prize in Physics for his work on theoretical physics in 1921? Answer: Albert Einstein, who is known as one of the best minds in physics.
- T or F? Yogurt is produced by the bacterial fermentation of milk. Answer: True.
- How many bones do sharks have in their bodies? Answer: None. Its body is made up of cartilage.
- The fear of what animal is known as 'arachnophobia'? Answer: Spiders.
- Does a human have feelings? Answer: Yes.
- In the beginning God created Heaven and _____? Answer: Earth.
- What is a lake? Answer: A body of water surrounded by land.
- Why do we dream? Answer: Its exact purpose is not known but dreaming may help us process our emotions.
- Are there other universes? Answer: Yes.
- About 97% of the Earth's water is in the _____. Answer: Ocean.
- Name at least two big lakes. Answers: Superior (North America), Victoria (Africa), Baikal (Russia), Tanganyika (Africa), Caspian Sea (Russia and Iran), Loch Ness (Scotland), Great Slave Lake (Canada), Lake Malawi (Africa).
- Name four oceans. Answers: Atlantic, Pacific, Indian and Arctic.
- Name three bodies of water. Answers: Seas, lakes, rivers, streams, glaciers.
- What is another word for water? Answers: H2O or Aqua.
- Why are oceans salty? Answers: Run off from the land, openings in the seafloor and rocks on land.
- Pure water is called? Answer: Distilled.
- Which ocean is not made up of salt water? Answer: Artic.
- What is a liquid? Answer: A substance that flows freely but is of constant volume.
- What is a solid? Answer: One of the four fundamental states of matter.
- What is a gas? Answer: One of the four fundamental states of matter.
- What is plasma? Answer: One of the four fundamental states of matter.
- What are two types of water? Answer: Storm water and wastewater.
- Name two types of bottled water. Answers: Fiji, Evian, Deer Park, Nestle, Dasani, Aquafina.
- Why is the sky blue? Answer: The sky looks blue but really it is made up of all colors of the rainbow.
- Can an airplane go in reverse? Answer: Yes. Airplane pilots usually only use this function for stopping once they land.
- Is there dry water? Answer: Yes, dry water is an unusual form of powdered liquid.
- Name three colors of a rainbow. Answers: Orange, yellow, green, blue, indigo, violet.
- Do aliens exist? Answer: We do not know. There is no scientific evidence that proves they exist.
- Most birds fly_____ for the winter. Answer: South.
- What is a mammal? Answer: A group of vertebrate animals in which the young are nourished with milk from the mammary glands of the mother.
- Are Humans mammals? Answer: Yes.

- Can a plane stop in the air? Answer: No. Planes need to keep moving to stay in the air.
- Gravity is a physical connection between space and _____ . Answer: Matter.
- _____ is anything you can touch physically. Answer: Matter.
- We use the word _____ to describe how much matter there is in something. Answer: Mass.
- The degree of compactness of a substance is _____. Answer: Density.
- The amount of space that a substance or object occupies is _____ . Answer: Volume.
- What country has the cleanest drinking water? Answer: Switzerland.
- The cleanest air in the world has been discovered in the _____. Answer: Antarctic Ocean.
- What was Einstein's most famous theory? Answer: The Theory of Relativity.
- What are worms? Answer: Many different distantly related animals that typically have a long cylindrical tube-like body, and has no limbs and eyes.
- What is a Vertebrate? Answer: An animal in a large group distinguished by the possession of a backbone or spinal column.
- What is an Invertebrate? Answer: An animal lacking a backbone.
- What are the major groups of Vertebrates? Answer: fish, amphibians, reptiles, mammals, and birds.
- What are two examples of Invertebrates? Answers: Poriferans (sponges) and Cnidarians (jellyfish).
- What is the Scientific Method? Answer: A method of research where a problem is identified, relevant data is gathered, a hypothesis is formulated from this data, and the hypothesis is empirically tested.
- What are the eight basic steps of the Scientific Method? Answer: See, ask a question, gather information, form a hypothesis, test the hypothesis, make a prediction, make a conclusion, report, and evaluate.
- What is the Solar System? Answer: A gravitationally bound system of the sun and the objects that orbit it, either directly or indirectly.
- Our _____ _____ consists of the sun, the planets, dozens of moons, comets and meteoroids. Answer: Solar System.
- How many planets are there in the Solar System? Answer: Eight.
- Name all eight planets. Answer: Mercury, Venus, Earth, Mars, Jupiter, Saturn, Uranus, and Neptune.
- All living things have _____ within their cells. Answer: DNA.
- Blood is always what color? Answer: Red.
- Trees give Earth its _____ . Answer: Oxygen.
- How long does it take our eyes to adapt to darkness? Answer: 20 to 30 minutes.
- Having a larger _____ size and volume is associated with better cognitive functioning and higher intelligence. Answer: Brain.
- Are bats blind? Answer: No. Bats have small eyes with extremely sensitive vision, which help them see in conditions we might consider pitch black.
- A reduced ability to distinguish between certain colors is _____ . Answer: Color blindness.
- Name at least three bodies of water. Answer: Sea, lake, ocean, stream, bay, lagoon, pond.

Music Vocabulary Words List

- **Accent**: when a specific note or phrase is emphasized with an increase in intensity.
- **Accordion:** a portable musical instrument with metal reeds blown by bellows, played by means of keys and buttons.
- **Adagio**: music should be played at a slower tempo.
- **Allegro**: music should be played at an upbeat and bright tempo.
- **Alto**: a range of pitches normally assigned to a singer in a choir. The alto range is below soprano but higher than the tenor range.
- **Andante**: a moderately slow tempo.
- **Arpeggio:** when a chord of notes is broken and played in sequence.
- **Bar:** a subsection of time that is defined by a time signature.
- **Bass:** a tone of low pitch.
- **Bass Drum:** a large two-headed drum that has a low booming sound.
- **Bassoon**: a bass instrument of the oboe family with a double reed.
- **Cadence**: a sequence of chords used to signify the end of a phrase.
- **Cadenza**: a moment in a musical piece where an instrumentalist or singer is given the opportunity to play a solo freely, and with artistic license to go outside of a rigid tempo or rhythm.
- **Canon**: when a melody is played by one instrument or group of instruments, and then repeated a certain number of bars later by another instrument to overlap the initial melody.
- **Cello**: a bass instrument of the violin family.
- **Clarinet**: a woodwind instrument with a single-reed mouthpiece, a straight cylindrical tube and flared bell.
- **Claves**: a percussion instrument consisting of a pair of short, wooden sticks.
- **Clef**: a symbol used at the beginning of a piece of sheet music to denote the note values on the staff.
- **Coda:** a symbol used in sheet music to denote where the final passage of a piece begins.
- **Congo**: a tall, narrow, single-headed drum from Cuba.
- **Crescendo**: a gradual increase in dynamic volume during a section of music.
- **Da Capo**: an instruction used in sheet music that tells the band or orchestra to restart the piece from the beginning.
- **Dal Segno**: an instruction used in sheet music that tells the band or orchestra to resume playing the piece from a different section, usually denoted by a star-like symbol or sign.
- **Decrescendo**: a movement in the piece calling for the music to gradually become softer.
- **Diminuendo**: a decrease in dynamic volume during a section of music.
- **Double Bass**: the largest and lowest-pitched bowed string instrument.
- **Dynamics**: how quietly or loudly a piece of music should be played.
- **Drumstick**: a type of percussion mallet used particularly for playing a snare drum.

- **Fermata**: a symbol used in sheet music to indicate that a note should be held longer than its standard duration.
- **Flat**: the relative tonal quality of a note.
- **Flute**: a wind instrument made from a tube with holes along it that are stopped by the fingers or keys, held vertically or horizontally so that the player's breath strikes a narrow edge.
- **Forte**: loud.
- **Fortepiano**: a dynamic instruction that tells an instrumentalist to initially play a note loudly and then quickly decay to a quiet sustained dynamic.
- **Fortissimo:** very loud.
- **Fortississimo**: very, very loud.
- **French Horn**: a brass instrument made of tubing wrapped into a coil with a flared bell.
- **Giocoso**: implies that the piece should be played in a fun and carefree manner, most often at a higher tempo.
- **Glissando**: instructs instrumentalists to slide in pitch from note to note, instead of accentuating each note.
- **Glockenspiel**: a pitched percussion instrument with metal bars that are struck by a hard mallet.
- **Guitar**: a flat-bodied stringed instrument with a long-fretted neck and usually six strings played with a pick or with the fingers.
- **Harp**: a plucked stringed instrument consisting of a resonator, an arched or angled neck that may be supported by a post.
- **Largo/Larghetto**: a large and slow-moving pace.
- **Legato**: to play lightly, usually at a quicker pace and in a light-hearted manner.
- **Leggero**: to connect each note smoothly without articulation between notes.
- **Mallet**: a device used by a percussion player to strike the instrument.
- **Maracas**: a rattle usually made from a gourd that is used as a percussion instrument.
- **Mezzo-forte**: medium loud.
- **Mezzo-piano**: medium soft.
- **Motif**: a specific melody or series of notes used in diverse ways throughout a piece of music or song.
- **Music**: an arrangement of sounds having, melody, rhythm, and usually harmony.
- **Natural**: a note that is neither sharp nor flat.
- **Nonet:** a group of nine musicians.
- **Oboe**: a woodwind instrument with a double-reed mouthpiece, a slender tubular body, and holes stopped by keys.
- **Organ**: a keyboard instrument operated by the player's hands and feet, in which pressurized air produces notes through a series of pipes organized in scalelike rows.
- **Ostinato**: a rhythmic pattern that repeats throughout a piece of music.
- **Pan**: refers to the stereo direction of the audio signal.
- **Pianissimo**: a dynamic instruction in music that tells musicians to play very softly or quieter.

- **Piano**: a passage marked to be performed softly; large keyboard musical instrument with a wooden case enclosing a soundboard and metal strings, which are struck by hammers when the keys are depressed.
- **Piccolo**: a small flute sounding an octave higher than the ordinary one.
- **Pizzicato**: instructs string sections to pluck their instruments instead of bowing them.
- **Pluck**: to cause the strings on a stringed instrument to vibrate by picking or pulling them with fingers or a pick.
- **Quarter Tone**: a musical interval that is half the value of a semitone and a quarter of the value of a whole tone.
- **Quintuplet**: a rhythmic notation that instructs players to play five notes in the space a quarter note uses.
- **Recorders**: a wooden or plastic musical instrument in the shape of a pipe, usually played by blowing into a mouthpiece and covering and uncovering the holes with your fingers.
- **Rhapsody**: a one-movement piece of music that explores multiple free-flowing sections that do not necessarily relate to one another.
- **Ritardando**: a music instruction that requires musicians to gradually slow down in tempo.
- **Rondo**: a type of orchestral form or song structure.
- **Saxophone**: a type of single-reed woodwind instrument with a conical body. Usually made of brass.
- **Scherzo**: refers to a short orchestral piece of music.
- **Sforzando**: a dynamic instruction that requires players to play a note abruptly and loudly.
- **Sharp**: a sharp musical note is a semitone higher in intonation than the same natural note.
- **Snare Drum**: a percussion instrument that produces a sharp staccato sound when the head is struck with a drumstick.
- **Soprano**: a range of pitches in the highest register of tones.
- **Sostenuto**: musical passages that require musicians to play each note beyond its normal value.
- **Staccato**: notes that are played much shorter than normal value.
- **Tambourine**: a percussion instrument with small metal disks in slots around the edge, played by being shaken or hit with the hand.
- **Tempo**: the pace or speed at which a piece of music is played.
- **Tenor**: a range of notes between alto and bass.
- **Timpani**: these are kettledrums.
- **Tremolo**: an effect musicians can put on a sustained note to create a trembling sound.
- **Triangle**: a percussion instrument consisting of a steel rod bent into a triangle with one corner left open.
- **Trill**: an instruction to sustain rapid alternation between two different pitches.
- **Trombone**: a brass instrument worked by lip vibrations against a cupped mouthpiece. It has an extendable slide that can increase the length of the instrument's tubing.
- **Trumpet**: a brass musical instrument with a flared bell and a bright, penetrating tone.

- **Tuba**: a large brass wind instrument of bass pitch, with three to six valves and a broad bell typically facing upward.
- **Vibrato**: an effect where the pitch of a note is subtly moved up and down to create a vibrating effect.
- **Viola**: an instrument of the violin family, larger than the violin and tuned a fifth lower.
- **Violin**: a stringed musical instrument of treble pitch.
- **Vivace**: a fast tempo, louder dynamic and bright playing.
- **Xylophone**: a percussion instrument that consists of wooden bars struck by mallets.

Math Vocabulary Words List

- **Absolute Value**: always a positive number.
- **Add**: to bring two or more numbers together to make a new total.
- **Addend**: a number involved in an addition problem.
- **Addition**: the action or process of adding something to something else.
- **Algebra**: the branch of mathematics that substitutes letters for numbers to solve for unknown values.
- **Angle**: two rays sharing the same endpoint.
- **Answer**: a solution to a problem, especially in mathematics.
- **Area**: the two-dimensional space taken up by an object or shape, given in square units.
- **Arithmetic**: a science that deals with the addition, subtraction, multiplication, and division of numbers.
- **Array**: a set of numbers or objects that follow a specific pattern.
- **Axis**: a line where a curve or figure is drawn, measured, and rotated.
- **Binary**: relating to, composed of, or involving two things.
- **Calculate**: to determine the amount or number of something.
- **Calculus**: the study of motion in which changing values are studied.
- **Capacity**: the volume of substance that a container will hold.
- **Circle**: a plane figure bounded by one curved line and is equal.
- **Circumference**: the length around a circle.
- **Compass**: a drawing instrument used for drawing circles and arcs.
- **Composite Number**: a positive integer with at least one factor aside from its own.
- **Cone**: a three-dimensional shape with only one vertex and a circular base.
- **Congruent**: objects and figures that have the same size and shape.
- **Coordinates**: two numbers, or sometimes a letter and a number, that locate a specific point on a grid.
- **Cosine**: the ratio of the length of the side adjacent to the angle to the length of the hypotenuse.
- **Counting**: the process of determining the number of elements of a finite set of objects.
- **Cube**: a symmetrical three-dimensional shape contained by six equal squares.
- **Curve**: an abstract term used to describe the path of a continuously moving point.
- **Cylinder**: a three-dimensional shape consisting of two circle bases connected by a curved type.
- **Decimal**: a real number on the base of a fraction.
- **Degree**: a unit in any of various scales of temperature, intensity, or hardness.
- **Denominator**: the number below the line in a common fraction.
- **Diameter**: a line that passes through the center of a circle and divides it in half.
- **Difference**: the answer to a subtraction problem, in which one number is taken from another.
- **Divide**: to split into equal parts.

- **Ellipse**: a closed curve, which is the intersection of a right circular cone.
- **Equal**: the same amount or value.
- **Equation**: a statement that illustrates equality between two expressions by joining them with an equals sign.
- **Equilateral**: having all sides of the same length.
- **Even Number**: a number that is divisible by 2.
- **Exponent**: the number that repeats the multiplication of a term, shown as superscript above the term.
- **Expression**: mathematical statements that have a minimum of two terms containing numbers or variables, connected by an operator in between.
- **Factor**: a number that divides into another number exactly.
- **Finite**: not infinite; has an end.
- **Focus**: a point used to construct a conic section.
- **Formula**: a rule that numerically describes the relationship between two or more variables.
- **Fraction**: a quantity that is not whole and contains a numerator and denominator.
- **Geometry**: the study of lines, angles, shapes, and their properties.
- **Graph**: a picture designed to express words.
- **Greater Than**: denoting one value to be greater than the other.
- **Half**: one or two equal parts into which something can be divided.
- **Hyperbola**: a type of conic section or symmetrical open curve.
- **Hypotenuse**: the longest side of a right-angled triangle.
- **Identity**: an equation that is true for variables of any value.
- **Inequality**: difference in size.
- **Integers**: all whole numbers, positive or negative, including 0.
- **Intersection**: a point or line common to lines or surfaces that intersect.
- **Inverse**: the opposite of another operation.
- **Irrational Number**: a number that cannot be represented as a decimal or fraction.
- **Isosceles**: a polygon with two sides of equal length.
- **Kilo**: a metric measure of mass.
- **Less Than**: denoting one value to be lesser than the other.
- **Line**: a long narrow mark or band.
- **Linear**: arranged in or extending along a straight line.
- **Logic**: sound reasoning and the formal laws of reasoning.
- **Long Division**: a method used for dividing large numbers into groups or parts.
- **Math**: the overall group of sciences that study numbers, shapes, and their relationships.
- **Mathematician**: an expert in or student in math.
- **Mathematics**: the study of such topics as quantity, structure, space, and change.
- **Mean**: this is the same as the average.
- **Median**: this is the middle value.

- **Minus**: a sign used in mathematics to indicate subtraction.
- **Multiple**: the product of that number and any other whole number.
- **Multiply**: the act or process of multiplying; the state of being multiplied.
- **Nano**: a unit prefix meaning one billionth.
- **Negative**: a real quantity having a value less than 0.
- **Null**: zero or none.
- **Number**: a mathematical object used to count, measure, and label.
- **Number Line**: a line on which numbers are marked at intervals used to illustrate simple numerical operations.
- **Numeral**: figures or symbols that denote a number.
- **Numerator**: the top number in a fraction.
- **Numerical**: relating to or expressed as a number or numbers.
- **Obtuse**: more than 90 degrees and less than 180 degrees.
- **Octagon**: a polygon with eight sides.
- **Odd Number**: a whole number that is not divided by 2.
- **Operation**: refers to addition, subtraction, multiplication, or division.
- **Parabola**: an open curve whose points are equidistant from a fixed point called the focus, and a fixed straight line called the directrix.
- **Parallel**: lines, planes, surfaces, or objects side by side and having the same distance continuously between them.
- **Parallelogram**: a quadrilateral with two sets of opposite sides that are parallel.
- **Percent**: one part in every hundred.
- **Perimeter**: the distance around the outside of a shape.
- **Perpendicular**: lines that intersect to form right angles.
- **Pi**: used to represent the ratio of a circumference of a circle to its diameter.
- **Plane**: when a set of points join together to form a flat surface that extends in all directions.
- **Plot**: graphical technique for representing a data sheet.
- **Plus**: a symbol that means addition or 'to add.'
- **Point**: a location represented by a dot.
- **Polygon**: a closed plane figure bounded by straight lines.
- **Polyhedron**: a solid figure with many plane faces, typically more than six.
- **Polynomial**: an expression of more than two algebraic terms.
- **Prime Numbers**: integers greater than 1 that are only divisible by themselves and 1.
- **Product**: the result of two or more numbers when multiplied together.
- **Proof**: an inferential argument for a mathematical statement.
- **Protractor**: a semi-circle device used for measuring angles.
- **Quadrilateral**: a four-sided polygon.
- **Quotient**: the solution to a division problem.

- **Radian**: a unit of measurement, equal to an angle at the center of a circle whose arc is equal in length to the radius.
- **Radius**: a distance found by measuring a line segment extending from the center of a circle to any point on the circle.
- **Rational Number**: a number that can be represented as the quotient of two integers.
- **Ray**: a straight line with only one endpoint that extends infinitely.
- **Real Number**: a quantity that can be expressed as an infinite decimal expansion.
- **Rectangle**: a parallelogram with four right angles.
- **Remainder**: the number left over when a quantity cannot be divided evenly.
- **Rhombus**: a parallelogram with four sides of equal length and no right angles.
- **Right Angle**: an angle equal to 90 degrees.
- **Scientific Notation**: this is a way of expressing numbers that are too large or too small to be conveniently written in decimal form.
- **Series**: a description of the operation of adding infinitely many quantities, one after the other.
- **Set**: a collection of elements.
- **Skip Counting**: method of counting forward by numbers other than 1.
- **Slope**: shows the steepness or incline of a line and is determined by comparing the positions of two points on the line.
- **Solve**: to find an answer to.
- **Sphere**: an object having a round solid shape; a ball or globe.
- **Square**: a plane figure with four equal straight sides and four right angles.
- **Stem and leaf**: a graphic system used to organize and compare data.
- **Square Root**: a value that, when multiplied by itself, gives the number.
- **Subtraction**: the operation of finding the difference between two numbers or quantities by taking one away from the other.
- **Subtrahend**: a quantity of numbers to be subtracted from another.
- **Sum**: the result of an addition operation.
- **Symbol**: a figure or a combination of figures that is used to represent a mathematical object.
- **Symmetry**: two halves that match perfectly and are identical across an axis.
- **Tangent**: a straight line touching a curve from only one point.
- **Tessellation**: congruent plane of figures or shapes that cover a plane completely without overlapping.
- **Term**: a piece of an algebraic equation.
- **Transversal**: a line that crosses or intersects two or more lines.
- **Trapezoid**: a quadrilateral with exactly one pair of parallel sides.
- **Triangle**: a three-sided polygon.
- **Trinomial**: a polynomial with three terms.
- **Unit**: a standard quantity used in measurement.
- **Variable**: a letter used to represent a numerical value in equations and expressions.

- **Venn diagram**: a diagram usually shown as two overlapping circles and is used to compare two sets.
- **Vertex**: the point of intersection between two or more rays, often called a corner.
- **Volume**: a unit of measurement describing how much space a substance occupies, or the capacity of a container.
- **Weight**: the measurement of how heavy something is.
- **Whole Number**: a positive integer.
- **X-axis**: the vertical axis in a coordinate plane.
- **X-coordinate**: the first element in an ordered pair.
- **Y-axis**: the axis on a plane cartesian coordinate system parallel to which ordinates are measured.
- **Yard**: a unit of linear measure equal to 3 feet.
- **Y-coordinate**: the second element in an ordered pair.
- **Zero**: the integer denoted by 0 that, when used as a counting number, means that no objects are present.

Science Vocabulary Words List

- **Accelerate**: a degree of quickness.
- **Analyze**: detail examination.
- **Astronomy**: the study of everything in the universe beyond Earth's atmosphere.
- **Astrophysics**: a branch of astronomy that studies the composition and origin of objects in outer space.
- **Atom**: the smallest unit into which matter can be divided without the release of electrically charged particles.
- **Bacteria**: a unicellular microorganism.
- **Barometer**: a pressure measuring instrument.
- **Beaker**: a lipped cylindrical glass container for measuring the volume.
- **Biochemistry**: the chemistry of the living cell.
- **Biologist**: a scientist who conducts research in biology.
- **Biology**: the science that deals with things that are alive, such as plants and animals.
- **Boiling**: this process occurs when matter reaches near its boiling point.
- **Botany**: the study of plants.
- **Brain**: a coordinating organ of the human body.
- **Brittle**: hard in nature but breaks without showing elasticity.
- **Bulb**: light-producing instrument.
- **Bunsen Burner**: a kind of gas burner used as laboratory equipment.
- **Burette**: a glass tube with a tap at one end. Made for delivering known volumes of a liquid.
- **Burning**: the process of combustion.
- **Cell**: the smallest unit that can live on its own; makes up the tissues of the body and all living things.
- **Chemical**: a compound or substance that has been purified or prepared, especially artificially.
- **Chemist**: an expert in chemistry.
- **Chemistry**: the branch of science that studies the properties, composition, and structure of elements and compounds.
- **Circuit**: a path that electricity follows when it flows.
- **Classify**: the categorization on a common base.
- **Climate**: the average weather in an area over a longer period of time.
- **Climatologist**: a scientist who studies weather and atmosphere patterns.
- **Climatology**: the study of the atmosphere and weather patterns over time.
- **Compare**: this is the estimation of things.
- **Conductor**: material that passes electricity.
- **Cuvette**: a straight-sided optically clear container for holding liquid samples in a spectrophotometer or other instrument.

- **Dark**: the absence of light.
- **Data**: facts and statistics collected for reference or analysis.
- **Datum**: a piece of information.
- **Deciduous**: a type of tree or shrub.
- **Decrease**: smaller or fewer in amount or size.
- **Directions**: a route of course.
- **Dissolve**: the action of a solid object breaking down in a liquid.
- **Electricity**: a form of energy resulting from the existence of charged particles.
- **Electrochemist**: chemist who specializes in electrons, voltage, chemicals, matter, and energy.
- **Electrochemistry**: the branch of chemistry that deals with the relations between electrical and chemical phenomena.
- **Element**: a part or aspect of something abstract.
- **Energy**: the capacity for doing work; how things change or move.
- **Entomology**: the branch of zoology concerned with the study of insects.
- **Expand**: to make larger.
- **Experiment**: a scientific procedure undertaken to make a discovery, test a hypothesis, or demonstrate a known fact.
- **Fact**: any true information.
- **Fission**: the process of separation.
- **Flask**: a container for liquids.
- **Fossil**: the remains or impressions of a prehistoric organism.
- **Friction**: resistance due to movement.
- **Funnel**: a tube or pipe that is wide at the top and narrow at the bottom. Used for guiding liquid or powder into a small opening.
- **Fusion**: the process of joining things.
- **Gas**: state of matter that can expand freely.
- **Genetics**: the study of heredity and the variation of inherited characteristics.
- **Geologist**: a scientist who studies the solid, liquid and gaseous matter of Earth and other planets.
- **Geology**: the science that studies the Earth's physical structure and substance, its history, and the processes that act on it.
- **Geophysics**: the physics of the Earth.
- **Glassware**: ornaments and articles made from glass.
- **Graduated Cylinder**: long, slender vessels used for measuring the volumes of liquids.
- **Graph**: a representation of data.
- **Gravity**: the force that attracts mass.
- **Herpetology**: the branch of zoology concerned with the study of amphibians.
- **Humidity**: a quantity expressing water vapor's amount.
- **Hygrometer**: humidity measuring instrument.

- **Hypothesis**: a proposed explanation.
- **Ichthyology**: the branch of zoology devoted to the study of fish.
- **Immunology**: a branch of zoology that covers the study of immune systems in all organisms.
- **Increase**: greater in amount or size.
- **Inference**: a conclusion.
- **Insulator**: a substance resistant to heat and electricity.
- **Investigation**: an action of finding facts.
- **Kinetic**: movement.
- **Lab**: a place equipped for experimental study in a science or for testing and analysis.
- **Laboratory**: a room or building equipped for scientific experiments.
- **Laws**: a system of specific rules.
- **Lepidoptery**: a branch of entomology concerning the scientific study of moths and butterflies.
- **Light**: something that makes vision possible.
- **Magnetic**: having magnetic properties.
- **Magnetism**: the force exerted by magnets when they attract or repel each other.
- **Magnify**: making objects larger in appearance by optical devices.
- **Mass**: the amount of matter present in any object or body.
- **Matter**: anything that takes up space and can be weighed.
- **Measure**: finding the length or capacity of something using a rule or standard.
- **Measurement**: a collection of quantitative or numerical data that describes a property of an object or event.
- **Mechanics**: the study of motion and force by applying mathematics.
- **Melting**: the process of liquefaction by heating.
- **Meteorologist**: an expert in the study of meteorology; a weather forecaster.
- **Meteorology**: the science dealing with the atmosphere and its phenomena.
- **Microbiologist**: a scientist who studies microscopic life forms and processes.
- **Microbiology**: the branch of science that deals with microorganisms.
- **Microscope**: an optical instrument used for viewing small objects.
- **Mineral**: naturally, existing inorganic substance.
- **Mineralogy**: the scientific study of minerals.
- **Mixture**: a combination of different things.
- **Molecule**: the smallest particle of a substance that has all of the physical and chemical properties of that substance.
- **Motion**: the phenomenon in which an object changes its position over time.
- **Nectar**: juicy fluid within flowers.
- **Neutron**: Sub-particle of an atom.
- **Observatory**: a building or place given over to or equipped for observation of natural phenomena.
- **Observation**: the action or process of observing something.

- **Observe**: the act of or the power to see or take notice of something.
- **Opaque**: difficult to see.
- **Organism**: an individual animal, plant, or single-celled life form.
- **Ornithology**: a branch of zoology dealing with birds.
- **Paleontology**: the study of the history of life on Earth based on fossils.
- **Particle**: a minute quantity or fragment.
- **Petri Dish**: a small shallow dish of thin glass or plastic with a loose cover, used especially for cultures.
- **Phase**: a region of space.
- **Physical Science**: the study of the inorganic world.
- **Physics**: the study of matter and energy.
- **Pipette**: a slender pipe or tube where insignificant amounts of liquids are taken up by suction for measuring.
- **Pitch**: the intensity of a sound.
- **Practical**: learning by doing.
- **Prediction**: a forecast.
- **Pressure**: force per unit area.
- **Prey**: an animal that is hunted for food by another animal.
- **Proton**: a constituent of an atom.
- **Quantum**: quantity of energy.
- **Quantum Mechanics**: science dealing with the behavior of matter and light on the atomic and subatomic scale.
- **Radiology**: the science dealing with x-rays and other high energy radiation.
- **Radiologist**: a person who uses x-rays or other high-energy radiation.
- **Reaction**: any chemical process that causes change.
- **Reflection**: the turning of light or energy from any surface.
- **Research**: a careful and detailed study into a specific problem, concern, or issue using the scientific method.
- **Results**: findings after an investigation.
- **Retort**: to reply to.
- **Rusting**: a chemical process that spoils iron.
- **Scale**: the range, mass or volume of a chemical reaction or process.
- **Science**: knowledge about the natural world that is based on facts learned through experiments and observation.
- **Scientist**: a person who is studying or has expert knowledge of one or more of the natural or physical sciences.
- **Seismology**: the science focused on earthquakes and artificially produced vibrations of the Earth.
- **Separate**: division by physical methods.
- **Shadow**: a dark area.
- **Solubility**: the ability of a substance to become dissolved.

- **Sun**: a medium-sized, main sequence star located in a spiral arm of the Milky Way galaxy, orbited by all the planets and other bodies in our solar system, and supplying heat to the Earth.
- **Telescope**: device used to form magnified images of distant objects.
- **Temperature**: the measure of hotness or coldness expressed in terms of any several scales, including Fahrenheit and Celsius.
- **Test Tube**: a thin glass tube closed at one end, used to hold tiny amounts of material for laboratory testing and experiments.
- **Theory**: an explanation for why things work or how things happen.
- **Thermometer**: a device that measures temperature or a temperature gradient.
- **Tissue**: a group of cells that have a similar structure and that function together as a unit.
- **Transparent**: clear.
- **Variable**: something that changes or can be changed.
- **Virologist**: a medical doctor that oversees the diagnosis, management, and prevention of infection.
- **Virology**: the study of viruses.
- **Volcano**: an opening in the Earth's crust through which lava, volcanic ash, and gasses escape.
- **Volcanologist**: a geologist who focuses on understanding the formation and eruptive activity of volcanoes.
- **Volcanology**: the study of volcanoes.
- **Volume**: the three-dimensional space that is occupied by an object.
- **Volumetric Flask**: a flask for use in a volumetric analysis that contains a specific volume when filled to an indicated level.
- **Watch Glass**: a circular concave piece of glass used in chemistry as a surface to evaporate a liquid.
- **Weather**: the state of the atmosphere at a place and time regarding heat, dryness, sunshine, wind, and rain.
- **Weight**: the amount or quantity of heaviness or mass.
- **Zoology**: a branch of biology that studies the members of the animal kingdom and animal life in general.

Health and Fitness Vocabulary Words List

- **Agility**: the ability to move quickly and easily.
- **Aerobics**: vigorous exercises designed to strengthen the heart and lungs.
- **Arrhythmia**: improper beating of the heart, whether irregular, too fast or too slow.
- **Balance**: an even distribution of weight enabling someone or something to remain upright and steady.
- **Body Composition**: a ratio of fat, muscle, bone, and other body tissues.
- **Cardiac Arrest**: sudden unexpected loss of heart function, breathing, and consciousness.
- **Cardio**: relating to the heart.
- **Cardiovascular**: relating to the heart and blood vessels.
- **Cardiovascular Disease**: heart conditions that include diseased blood vessels, structural problems, and blood clots.
- **Cardiovascular System**: the heart, blood vessels, and blood.
- **Congenital Heart Disease**: an abnormality in the heart that develops before birth.
- **Coronary Artery Disease**: damage or disease in the heart's major blood vessels.
- **Endurance**: the ability to perform vigorous activity without getting overly tired.
- **Exercise**: activity requiring physical effort, carried out to sustain or improve health and fitness.
- **Fitness**: being in good physical shape or being suitable for a specific task or purpose.
- **Flexibility**: the quality of bending without breaking.
- **High Blood Pressure**: the quality in which the force of the blood against the artery walls is too high.
- **Muscle**: the tissue of the body which primarily functions as a source of power.
- **Muscle Endurance**: ability to use muscles for an extended period of time.
- **Nutrition**: the process of providing or obtaining the food necessary for health and growth.
- **Nutritionist**: a person who studies or is an expert in nutrition.
- **Peripheral Artery Disease**: a circulatory condition in which narrowed blood vessels reduce blood flow to the limbs.
- **Physical Fitness**: capacity of the whole body to function at optimum efficiency.
- **Protein**: a naturally occurring substance that consists of amino acid residues joined by peptide bonds.
- **Robustness**: the quality or condition of being strong and in good condition.
- **Shape**: the condition or state of someone or something.
- **Stamina**: the ability to sustain prolonged physical or mental effort.
- **Static Stretching**: slowly moving a muscle to its stretching point and holding it for 15 seconds.
- **Strength**: the ability of your muscles to exert a force.
- **Stroke**: damage to the brain from an interruption of its blood supply.

- **Toughness**: the state of being strong enough to withstand adverse conditions or rough handling.
- **Vegetable**: a plant used as food.
- **Verdure**: lush green vegetation.
- **Vigor**: physical strength and good health.
- **Vitamins**: any group of organic compounds which are essential for normal growth and nutrition, and are required in small quantities in the diet because they cannot be synthesized by the body.
- **Well-being**: the state of being comfortable; healthy, or happy.
- **Wellness**: the state of being in especially good health.
- **Zumba**: an aerobic fitness program featuring movements inspired by various styles of Latin American dance and performed primarily to Latin American dance music.

Printed in the United States
by Baker & Taylor Publisher Services